Beatrice Hill Tinsley
Queen of the Cosmos

Maria Gill
Illustrated by Alistair Hughes

upstart press

"YOU CAN DO ANYTHING
IF YOU PUT YOUR
MIND TO IT"
BEATRICE HILL TINSLEY

A gifted child brought up in Christchurch and then New Plymouth became a pioneer in astronomy and cosmology. Beatrice Hill Tinsley challenged long-held ideas about how old stars were, how far away galaxies were, and whether the universe was expanding or would one day expire. Her findings had a lasting effect on cosmology and astronomy.

In 1981, New Zealand's Prime Minister Bill Rowling said that New Zealand has given the world three great scientists: Ernest Rutherford, William Pickering, and Beatrice Hill Tinsley.

One snowy night during World War II, a German pilot dropped two bombs near the Hill family home in Chester, England. The explosion gave pregnant Jean Hill a terrible fright. Soon after, baby Beatrice arrived seven weeks early on 27 January 1941.

The doctor told the new parents, "The baby's room needs to be as warm as a hot summer day to help her survive." He left, shaking his head, knowing rationed coal was hard to get. The neighbours spread the word that the Hill family needed fuel for their fireplace. Soon, they had plenty of coal to heat the nursery throughout the freezing winter.

The Hill family moved away from the nightly terrors of bombs whistling down during the war. In the country air, Beatrice grew into a healthy young child.

Her favourite game was pretending she was a radio announcer. "The next item will be a violin solo. The violinist will be Beatrice Hill. The pianist is someone no one has ever heard of."

But it was difficult living in England after the war. The weather was miserable; the family caught whooping cough and chicken pox; and they still had to use ration cards to buy food. When a family friend offered Beatrice's dad a job in Christchurch, he saw a chance for a better life. The family boarded RMS *Akaroa* in July 1946 for the other side of the world.

By the time the family moved from Christchurch to New Plymouth, Beatrice was nine years old and excelling at school. Each morning she practised the piano and violin. After school, she scribbled notes and equations to remember what her primary teachers had taught her.

Over the weekends, she played imaginary games with her younger sister Theodora and taught their corgi circus tricks. She made Bruce, her teddy, a harp-like musical instrument, so he could learn music with her.

At high school, Beatrice read about astronomy while browsing the *Junior Oxford Encyclopaedia*. What a fascinating subject, she thought. She joined the local astronomy club and enrolled in physics. Beatrice asked her teacher, "Can I read those books on the top shelf?"

Though Beatrice had just started learning physics, she was already working above her age. The teacher let her borrow the advanced physics books. From then on, Beatrice studied by herself.

In her last year at high school, she topped all her subjects and won a scholarship to study at Canterbury University.

Whenever Beatrice looked up at the night sky, questions flooded her mind. How many stars are in the Milky Way? She was astounded to learn there were over a hundred billion stars and the Milky Way was only one of many galaxies.

How old is the universe? Where did it come from? Where is it going? She wanted to understand and know as much as possible about the cosmos.

Beatrice studied physics and mathematics at university, tutored other students, and played the violin in the National Youth Orchestra. While studying for her master's degree, she married Brian Tinsley, a fellow student.

In October 1963, twenty-two-year-old Beatrice and Brian moved to Dallas in the USA for his new job. She was told she could continue her studies in cosmology. But she soon discovered Texan universities nearby didn't teach it at her level. A frustrated Beatrice applied to study in a town over 600 kilometres away.

When the professor interviewed Beatrice for the doctorate, the highest degree you can study, he thought she was just a young high school graduate. "Cosmology is a difficult subject, you know."

"I do know," said Beatrice. She'd never had less than an 'A' grade ever.

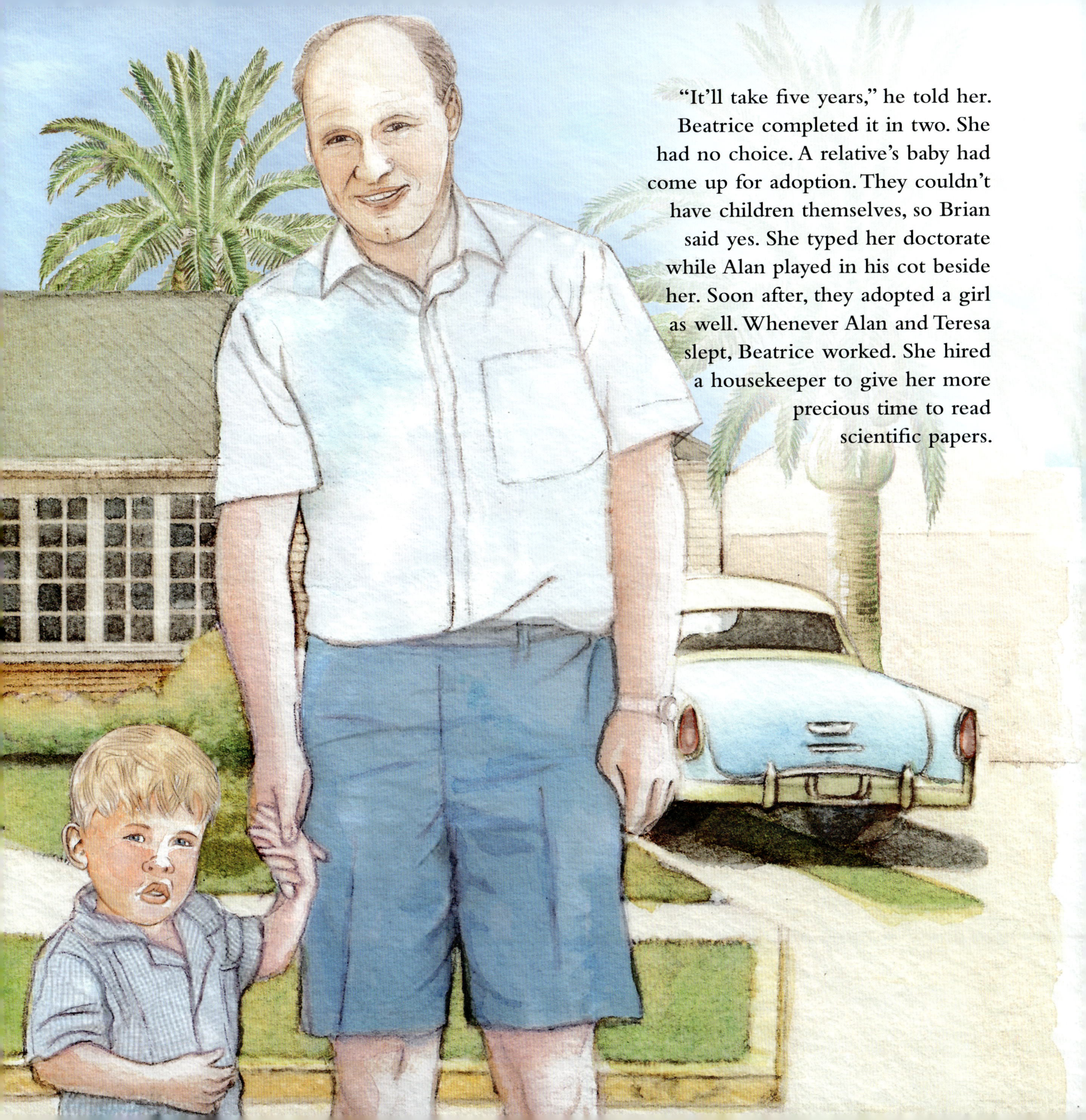

"It'll take five years," he told her. Beatrice completed it in two. She had no choice. A relative's baby had come up for adoption. They couldn't have children themselves, so Brian said yes. She typed her doctorate while Alan played in his cot beside her. Soon after, they adopted a girl as well. Whenever Alan and Teresa slept, Beatrice worked. She hired a housekeeper to give her more precious time to read scientific papers.

Scientists around the world heard about this brilliant New Zealand cosmologist. They invited her to speak at their conferences. She told Brian the exciting news. He was thrilled for her. "But who will look after the children?" Beatrice organised babysitters. What she really wanted, though, was to move to a state where they could both do their research. But Brian was reluctant. His job enabled him to travel all over the world. He didn't want to move to an uncertain future.

After living ten years in Dallas with no hope of continuing her important work, Beatrice decided to leave her husband. "The kids can come with me," she said, thinking he was away too much to look after them. A hurt Brian replied, "I will fight you in every court in America!"

Beatrice didn't want her children to go through that stress. Distraught, she left them with Brian. The children visited Beatrice in the holidays. She missed them every day and wrote them many letters.

Beatrice made the most of her time, writing over 100 scientific papers. She became the first female professor of astronomy at Yale University, teaching the next generation of scientists. And she worked with top cosmologists from around the world.

Beatrice theorised that the further away a galaxy was, the longer ago its light must have left to reach Earth. Galaxies dim with age as the stars inside them changed. This knowledge helped her answer the biggest question in cosmology. Would the universe keep expanding forever or would it one day collapse? She calculated it would keep expanding at what scientists now know to be an ever-increasing speed.

The most well-known astronomer at the time disagreed with her theory at first, but other astronomers and cosmologists quickly accepted it. Over her short career, Beatrice became known as the leading expert on the evolution of galaxies.

When she was thirty-seven years old, Beatrice noticed a black mole on the side of her leg.

"It's melanoma, the deadliest form of skin cancer," the doctor told Beatrice after tests. She hoped the radiation treatment would cure her. But the cancer returned. In the short time she might have left, Beatrice organised her life. In between visiting specialists and stays in hospital, she helped her eleven-year-old daughter Teresa, who now lived with her, settle into a new school.

While Beatrice lay dying in the hospital, she continued writing scientific articles. In her last letter to her father, she wrote: "I've been one of the luckiest people ... to realise my lifelong ambitions and far more so than I could have hoped..."

The Difficulties Beatrice Faced

Back then, people believed women should stay home, do the housework, and mind the children. Beatrice defied those beliefs and accomplished astounding feats.

Universities didn't employ wives of professors, however outstanding they were in their work. They didn't want to be seen as showing favouritism.

Beatrice looked at the stars to find answers. Sadly, one of those stars – our sun – would give her a skin cancer called melanoma. If she had gone to the doctors when she first noticed the dark mole and had treatment, the cancer might not have spread. Had people back then known how important it is to wear sunscreen when out in the sun, she might never have had melanoma and still be alive today.

Beatrice lived every day as if she was on borrowed time. From her premature birth, to dying young at 40 years old. During that time, she was determined to learn and excel at whatever she tried and then, as a scientist, further the world's knowledge of cosmology. And she achieved it in the short time she had on Earth. If she had lived, she would have made many more exciting discoveries.

SCIENTISTS CONSIDERED BEATRICE HILL TINSLEY THE QUEEN OF COSMOLOGY.

Her questioning mind and creative approach to finding new theories had a huge impact on cosmology. No one had thought to sift through research in different fields to find links and look back at telescopic images to provide insights. She then proved her theories with precise mathematical calculations and computer modelling.

Her contributions were so significant that New Zealand and the USA wanted her memory to live on and named these after her:

- **Two New Zealand postage stamps**
- **A street in Albany, Auckland**
- **A laboratory at Wellington East Girls' College**
- **A mountain in Fiordland**
- **A science precinct at the University of Canterbury**
- **Asteroid 3087**
- **The Beatrice Tinsley Fund helps outstanding students in the USA**
- **The Beatrice M. Tinsley Prize is awarded to scientists who have made an outstanding contribution to astronomy**
- **A play called *Bright Star* based on her life**

Beatrice Tinsley's timeline

1941
27 January - born in Chester, UK

1946
Boarded RMS *Akaroa* and travelled to Christchurch, NZ

1950
Family moved to New Plymouth

1953
Joined astronomy club at New Plymouth Girls' High School

1967
Awarded PhD (doctorate) from Austin University

1968
Adopted Teresa Tinsley

1974
Divorced Brian Tinsley and pursued career

1975
Worked at Yale University

1978
First female professor of astronomy at Yale. Diagnosed with melanoma

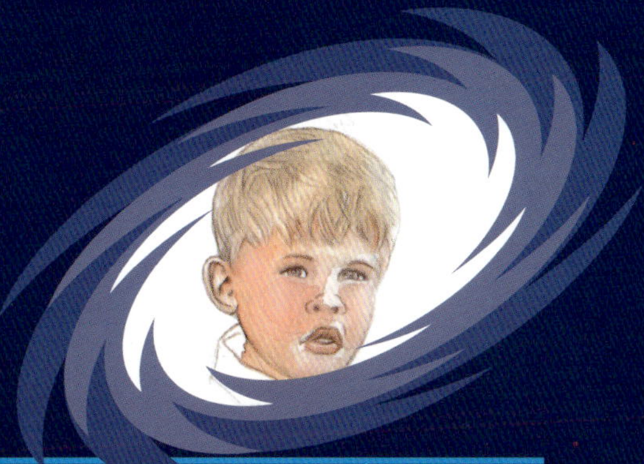

1957
Won Junior University scholarship to go to Canterbury University

1961
Married Brian Tinsley. Awarded Bachelor of Science

1963
Moved to Dallas, Texas. Awarded Master of Science

1966
Adopted Alan Tinsley

1981
Died aged 40 years old. Asteroid 3087 named in honour of Beatrice

1986
The American Astronomical Society established the Beatrice M. Tinsley Prize

2009
University of Canterbury formed the Beatrice Tinsley Institute

2016
The New Zealand Association of Scientists renamed the Research Medal the Hill Tinsley Medal

2019
The Beatrice Tinsley building opened at the University of Canterbury

GLOSSARY

ASTRONOMY
SCIENTIFIC STUDY OF ALL OBJECTS IN SPACE.

COSMOLOGY
SCIENTIFIC STUDY OF THE ORIGIN AND DEVELOPMENT OF THE UNIVERSE.

COSMOS
ANOTHER NAME FOR UNIVERSE.

EVOLUTION
GRADUAL DEVELOPMENT OF SOMETHING.

GALAXY
MASSIVE COLLECTION OF BILLIONS OF STARS, GAS, DUST, AND A SOLAR SYSTEM HELD TOGETHER BY GRAVITY.

MELANOMA
AGGRESSIVE SKIN CANCER.

MILKY WAY
LARGE SPIRAL GALAXY.

RADIATION
BEAM OF RADIATION (HIGH ENERGY RAYS) DIRECTED ONTO THE CANCER CELLS TO KILL IT.

RATION CARDS
CARD ISSUED BY THE GOVERNMENT TO ALLOW THE HOLDER TO OBTAIN FOOD THAT IS IN SHORT SUPPLY.

A catalogue record for this book is available from the National Library of New Zealand

ISBN 978-1-77694-005-9
An Upstart Press Book
Published in 2024 by Upstart Press Ltd
26 Greenpark Road, Penrose, Auckland 1061
New Zealand

Acknowledgements
I am grateful to Beatrice Hill Tinsley's family; Teresa Tinsley, Alan Tinsley, Theodora Lee-Smith; and friend and colleague Dr Richard Larson, for their fact checking of the manuscript.

Special thanks to the queen of my galaxy, and to Lily.

Design and Illustration by Alistair Hughes

Printed by Everbest Printing, China